AVIATION MAINTENANCE TECHNICIAN— GENERAL, AIRFRAME, AND POWERPLANT KNOWLEDGE TEST GUIDE

CONTENTS

INTRODUCTION

FAA-G-8082-3A, Aviation Maintenance Technician—General, Airframe, and Powerplant Knowledge Test Guide, provides information for preparing you to take one or all of the following knowledge tests. This document supersedes FAA-G-8082-3, Aviation Mechanic General, Airframe, and Powerplant Knowledge Test Guide, dated 19

TEST NAME	TEST CODE
Aviation Maintenance Technician—General	AMG
Aviation Maintenance Technician—Airframe	AMA
Aviation Maintenance Technician—Powerplant	AMP

The minimum passing score is 70 percent.

Federal Aviation Administration (FAA) airman knowledge tests are effective instruments for aviation safety and regulation measurement. However, these tests can only sample the vast amount of knowledge every aviation maintenance technician needs.

Comments may be e-mailed to AFS630Comments@faa.gov.

KNOWLEDGE TEST ELIGIBILITY REQUIREMENTS

The general qualifications for an aviation maintenance technician certificate require you to have a combination of experience, knowledge, and skill. If you are pursuing an aviation maintenance technician certificate with airframe and powerplant ratings, you should review the appropriate sections of Title 14 of the Code of Federal Regulations (14 CFR) part 65 for detailed information pertaining to eligibility requirements. Further information may be obtained from the nearest Flight Standards District Office (FSDO).

Before taking the certification knowledge and practical tests, you must meet the eligibility requirements. The determination of eligibility of applicants for the general, airframe, and powerplant tests is made on the basis of one of the following options:

1. **Civil and/or military experience.** (See 14 CFR Part 65, Certification: Airmen Other Than Flight Crewmembers, Subpart A—General and Subpart D— Mechanics.) If you believe you are qualified to exercise this option, you must have your experience evaluated and certified by an FAA aviation safety inspector (airworthiness). If the inspector determines that you have the required experience, two FAA Forms 8610-2, Airman Certificate and/or Rating Application, are completed. These forms are issued, and MUST be presented along with appropriate identification to take the corresponding knowledge tests. Your eligibility to test does not expire.

2. **Graduation from an FAA-certificated aviation maintenance technician school (AMTS).** Depending upon the testing facility affiliation, a graduation certificate, certificate of completion, or an FAA Form 8610-2, Airman Certificate and/or Rating Application (properly endorsed), is required, along with proper identification.

If you are taking the tests at a computer testing center and the practical testing is administered by a designated mechanic examiner (DME), and *both* are affiliated with the AMTS, a copy of the graduation certificate or certificate of completion (along with proper identification) may be all that you are required to present. In this case, the school, the testing center, the DME, and the local FSDO will all be involved and know what authorization is needed. On the other hand, if either or both the testing center and the DME are *not* affiliated with the AMTS, then FAA Form 8610-2 is required.

KNOWLEDGE AREAS ON THE TESTS

Aviation maintenance technician tests are comprehensive because they must test your knowledge in many subject areas. The subject areas for the tests are the same as the required AMTS curriculum subjects listed in 14 CFR part 147, Aviation Maintenance Technician Schools, appendices B, C, and D. However, the subject area titled "Unducted Fans" (in appendix D) is not a tested subject at this time. The terms used in 14 CFR part 147, appendices B, C, and D, are defined in 14 CFR part 147, appendix A.

DESCRIPTIONS OF THE TESTS

All test questions are the objective, multiple-choice type. Each question can be answered by the selection of a single response. Each test question is independent of other questions; therefore, a correct response to one does not depend upon, or influence, the correct response to another. **The minimum passing score is 70 percent.**

The Aviation Maintenance Technician—General test contains 60 questions, and you are allowed 2 hours to complete the test.

The Aviation Maintenance Technician—Airframe and Aviation Maintenance Technician— Powerplant tests contain 100 questions each, and you are allowed 2 hours to complete each test.

TEST REGISTRATION

The first step in taking a knowledge test is the registration process. You may either call one of the computer testing designees, (refer to the Computer Testing Designee section at the end of this document for 1-800 numbers) or simply use the walk-in basis. If you choose to use the 1-800 number to register, you will need to select a testing center, schedule a test date, and make financial arrangements for test payment. You may register for tests several weeks in advance, and you may cancel your appointment according to the CTD's cancellation policy. If you do not follow the CTD's cancellation policies, you could be subject to a cancellation fee.

The next step in taking a knowledge test is providing proper identification. An acceptable identification document includes a recent photograph, signature, and actual residential address, if different from the mailing address. This information may be presented in more than one form of identification.

Acceptable forms of identification include, but are not limited to, drivers' licenses, government identification cards, passports, alien residency (green) cards, and military identification cards. Other forms of identification that meet the requirements of this paragraph are acceptable. Some applicants may not possess the identification documentation described. In any case, you should always check with your local FSDO or IFO if you are unsure of the kind of identification to bring to the interview.

You also need to present authorization to test. Acceptable forms of authorization

aFAA

- Form 8610-2.
- A graduation certificate or certificate of completion to an affiliated testing center as previously explained.
- An original (not photocopy) failed Airman Knowledge Test Report, passing Airman Knowledge Test Report, or expired Airman Knowledge Test Report.

Before you take the actual test, you will have the option to take a sample test. The actual test is time limited; however, you should have sufficient time to complete and review your test.

TAKING THE TEST

Communication between individuals through the use of words is a complicated process. In addition to being an exercise in the application and use of aeronautical knowledge, a knowledge test is also an exercise in communication since it involves the use of the written language. Since the tests involve written rather than spoken words, communication between the test writer and the person being tested may become a difficult matter if care is not exercised by both parties. Consequently, considerable effort is expended to write each question in a clear, precise manner. Make sure you read the instructions given with the test, as well as the statements in each test item.

When taking a test, keep the following points in mind:

- Answer each question in accordance with the latest regulations and guidance publications.
- Read each question carefully before looking at the answer options. You should clearly understand the problem before attempting to solve it.
- After formulating an answer, determine which answer option corresponds with your answer. The answer you choose should completely resolve the problem.
- From the answer options given, it may appear that there is more than one possible answer; however, there is only one answer that is correct and complete. The other answers are either incomplete, erroneous, or derived from popular misconceptions.
- If a certain question is difficult for you, it is best to mark it for review and proceed to the next question. After you answer the less difficult questions, return to those you marked for review and answer them. The review marking procedure will be explained to you prior to starting the test. Although the computer should alert you to unanswered questions, make sure every question has an answer recorded. This procedure will enable you to use the

available time to maximum advantage.

- When solving a calculation problem, select the answer that most nearly matches your solution. The problem has been checked by various individuals and with different types of calculators; therefore, if you have solved it correctly, your answer will be closer to the correct answer than any of the other choices.

USE OF TEST AIDS AND MATERIALS

You may use aids, reference materials, and test materials within the guidelines listed below, if actual test questions or answers are not revealed. All models of aviation-oriented calculators may be used, including small electronic calculators that perform only arithmetic functions (add, subtract, multiply, and divide). Simple programmable memories, which allow addition to, subtraction from, or retrieval of one number from the memory, are permissible. Also, simple functions such as square root and percent keys are permissible.

The following guidelines apply:

1. You may use any reference materials provided with the test. In addition, you may use scales, straightedges, protractors, plotters, and electronic or mechanical calculators that are directly related to the test.

2. Manufacturer's permanently inscribed instructions on the front and back of such aids (e.g., formulas, conversions, and weight and balance formulas) are permissible.

3. Testing centers may provide a calculator to you and/or deny use of your personal calculator based on the following limitations:

   a. Prior to and upon completion of the test while in the presence of the proctor, you must actuate the ON/OFF switch and perform any other function that ensures erasure of any data stored in memory circuits.

   b. The use of electronic calculators incorporating permanent or continuous type memory circuits without erasure capability is prohibited. The proctor may refuse the use of your calculator when unable to determine the calculator's erasure capability.

   c. Printouts of data must be surrendered at the completion of the test if the calculator incorporates this design feature.

   d. The use of magnetic cards, magnetic tapes, modules, computer chips, or any other device upon which prewritten programs or information related to the test can be stored and retrieved is prohibited.

   e. You are not permitted to use any booklet or manual containing instructions related to use of test aids.

4. Dictionaries are not allowed in the testing area.

5. The proctor makes the final determination relating to test materials and personal possessions you may take into the testing area.

DYSLEXIC TESTING PROCEDURES

If you are a dyslexic applicant, you may request approval from the local Flight Standards District Office (FSDO) or International Field Office (IFO) to take an airman knowledge test using one of the three options listed here in preferential order.

Option 1. Use current testing facilities and procedures whenever possible. 4

Option 2. You may use a Franklin Speaking Wordmaster® to facilitate the testing process. The Wordmaster® is a self-contained electronic thesaurus that audibly pronounces typed-in words and presents them on a display screen. It has a built-in headphone jack for private listening. The headphone feature must be used during testing to avoid disturbing others.

Option 3. If you do not choose to use the first or second option, you may request a proctor to assist in reading specific words or terms from the test questions and supplement material. In the interest of preventing compromise of the testing process, the proctor must be someone who is non-aviation oriented. The proctor must provide reading assistance only, with no explanation of words or terms. When this option is requested, the FSDO or IFO inspector must contact the Airman Testing Standards Branch (AFS-630) for assistance in selecting the test site and proctor.

Prior to approval of any option, the FSDO or IFO inspector must advise you of the regulatory certification requirement of being able to read, write, speak, and understand the English language.

CHEATING OR OTHER UNAUTHORIZED CONDUCT

Computer testing centers are required to follow strict security procedures to avoid test compromise. These procedures are established by the FAA and are covered in FAA Order 8080.6, Conduct of Airman Knowledge Tests. The FAA has directed testing centers to terminate a test at any time a test proctor suspects a cheating incident has occurred. An FAA investigation will then be conducted. If the investigation determines that cheating or other unauthorized conduct has occurred, then any airman certificate or rating that you hold may be revoked, and you will be prohibited for 1 year from applying for or taking any test for a certificate or rating.

KNOWLEDGE TEST REPORTS

Upon completion of the knowledge test, you will receive your Airman Knowledge Test Report (with the testing center's embossed seal), which reflects your score.

The Airman Knowledge Test Report lists the learning statement codes for questions answered incorrectly. The total number of learning statement codes shown on the Airman Knowledge Test Report is not necessarily an indication of the total number of questions answered incorrectly.

The Learning Statement Reference Guide for Airman Knowledge Testing, found at www.faa.gov, contains the listings of reference materials, learning statement codes, and learning statements. The learning statement codes, as used in airman testing, refer to a measurable statement of knowledge that a student should be able to demonstrate following a defined element of training. You should match the codes on your Airman Knowledge Test Report to the codes in the Learning Statement Reference Guide to review your areas of deficiency.

The Airman Knowledge Test Report must be presented to the examiner prior to taking the practical test. During the oral portion of the practical test, the examiner is required to evaluate the noted areas of deficiency.

Should you require a duplicate Airman Knowledge Test Report due to loss or destruction of the original, send a signed request accompanied by a check or money order for $1 payable to the FAA. Your request should be sent to:

Federal Aviation Administration
Airmen Certification Branch, AFS-760
P.O. Box 25082
Oklahoma City, OK 73125

Airman Knowledge Test Reports are valid for the 24-calendar-month period preceding the month you complete the practical test. If the Airman Knowledge Test Report expires before completion of the practical test, you must retake the knowledge test.

RETESTING PROCEDURES

If you receive a grade lower than 70 percent and wish to retest, you must present the following to testing center personnel:

- You may retake the test after 30 days from the date your last test was taken by presenting your failed Airman Knowledge Test Report.
- You may retest sooner that 30 days if you present your failed Airman Knowledge Test Report and a signed statement from an airman holding the certificate and rating you seek certifying that you have been given additional instruction in each subject failed and that you are now ready for retesting.

If you decide to retake a test you passed in anticipation of a better score, you may retake the test after 30 days from the date your last test was taken. The FAA will not allow you to retake a passed test before the 30-day period has lapsed. Prior to retesting, you must give your current Airman Knowledge Test Report to the test proctor. The score from the last test taken will be the official score.

TRAINING AND TESTING PUBLICATIONS AND GENERAL INFORMATION

Most of the current Flight Standards Service airman training and testing publications can be obtained in electronic format from the FAA Web site, www.faa.gov. The training and testing publications and general information can be found on the opening page of that Web site under the Education and Research tab. If a publication is not available in electronic format, there are instructions for obtaining paper copies. Information found on the Web site includes the following:

- Advisory Circulars
- Airworthiness Directives
- Code of Federal Regulations
- Computer Testing Supplements
- Knowledge Test Centers
- Knowledge Test questions
- Knowledge Test statistics
- Learning Statement Reference Guide
- Practical Test Standards
- Training handbooks
- Type Certificate Data Sheets

Advisory Circulars

Advisory circulars (ACs) provide guidance and information on various subjects related to airman certification.

Airworthiness Directives

Airworthiness Directives (ADs) are notifications to aircraft owners of a known safety deficiency with a specific model of aircraft, engine, avionics, or other system.

Code of Federal Regulations

The portion of 14 CFR containing what was formerly known as the Federal Aviation Regulations can be found on the Web site. 14 CFR contains regulations designed to promote aviation safety, and govern all aviation activities in the United States.

Computer Testing Supplements

The knowledge testing supplements contain the graphics, legends, and maps that are needed to successfully respond to certain knowledge test items. These supplements will be provided by CTD test center personnel during the airman knowledge test.

Knowledge Test Centers

The Knowledge Test Centers portion of the Web site contains current listings of computer testing designees (CTDs) and other testing centers, and the registration telephone numbers to call to register for a test.

The following is a list of the computer testing designees authorized to give FAA airman knowledge tests. This list should be helpful in case you choose to register for a test or simply want more information.

- **Computer Assisted Testing Service (CATS)**
  1801 Murchison Drive, Suite 288
  Burlingame, CA 94010
  Applicant inquiry and test registration: 1-800-947-4228
  From outside the U.S. (650) 259-8550
- **LaserGrade Computer Testing**
  16821 SE McGillivray Blvd., Suite 201
  Vancouver, WA 98683
  Applicant inquiry and test registration: 1-800-211-2753 or 1-800-211-2754
  From outside the U.S. (360) 896-9111

Knowledge Test Questions

Sample questions are located in the Airman Knowledge Test Questions section of Web site and represent the types of questions included in the actual test banks. Practicing these questions will help you become familiar with similar questions on the airman knowledge tests. The knowledge test is not designed to intimidate any prospective airman; it is designed to measure understanding of the rules and regulations required to receive an FAA certificate.

Knowledge Test Statistics

Test statistics for all airman knowledge tests are contained in a series of tables organized by year and subject area. Individual tables are provided for the following subject areas: test volume, pass rates, average test scores, countries, regions, and district offices.

Learning Statement Reference Guide

Learning statement codes replace the old subject matter codes and are noted on the test report. They refer to measurable statements of knowledge that a student should be able to demonstrate following a defined element of training. The learning statement corresponding to the learning statement code on the test report can be located in the Learning Statement Reference Guide on the Web site.

Practical Test Standards

The practical test standards outline the knowledge and skill requirements for each airman certificate and rating. The references listed in each task of the practical test standards indicate the specific publications used to develop the skill standards. The ability to issue immediate changes prior to publishing revised printed copies ensures the practical test standards are always accurate and usable.

Training Handbooks

The training handbooks are the basic information sources an airman applicant should refer to when preparing for the knowledge and practical tests for a specific certificate or rating.

Type Certificate Data Sheets (TCDS)

A Type Certificate Data Sheet contains a formal description of an aircraft, engine, or propeller, including detailed specification of the type design and the information required for type certification.

SAMPLE TEST QUESTIONS AND ANSWERS

Aviation Maintenance Technician—General

1. **If the cross-sectional area of a given conductor is increased to four times its original value, and the length and temperature remain constant, the resistance of the conductor will be**

   A) one-fourth its original value.
   B) four times its original value.
   C) found by multiplying the original resistance by the percentage increase in cross-sectional area.

   Answer: A. Learning Statement: AMG031. FAA-H-8083-30, Aviation Maintenance Technician Handbook—General.

   One of the factors affecting the resistance of a conductor is cross-sectional area. Resistance varies inversely with the cross-sectional area of a conductor. If the cross-sectional area of a conductor is doubled, the resistance to current flow will be reduced by half (all other factors remaining unchanged).

2. **When making a forward weight and balance check to determine that the center of gravity (CG) will not exceed the forward limit during extreme conditions, the items of useful load which should be computed at their minimum weights are those located aft of the**

A) forward CG limit.

B) rearward CG limit.

C) empty weight CG.

Answer: A. Learning Statement: AMG002. FAA-H-8083-30, Aviation Maintenance Technician Handbook—General.

When making a forward weight and balance check, part of the information needed is the minimum weight of the items of useful load that are located aft of the forward CG limit.

3. What must a certificated mechanic with both airframe and powerplant ratings do prior to returning to service an aircraft on which he or she has performed and approved a 100-hour inspection?

A) Present his or her work and records to a mechanic holding an Inspection Authorization for final approval and release.

B) Make the proper entries in the appropriate aircraft maintenance record.

C) Notify the local FAA FSDO in writing of his or her intention to return the aircraft to service.

Answer: B. Learning Statement: AMG080. 14 CFR section 43.11(a).

The person approving for return to service an aircraft after any inspection shall make an entry containing the required information in the maintenance record.

Aviation Maintenance Technician—Airframe

1. Which of the following drill bit types work best when drilling an aramid fiber (Kevlar®) composite laminate?

A) Tool steel with standard grind.

B) Diamond dust coated.

C) Carbide W-Point.

Answer: C. Learning Statement: AMG020. AMR, Aircraft Maintenance and Repair.

Standard tool steels dull rapidly when drilling or trimming composite materials. If diamond-dust coated drills are used, the fibers will grab at the drill bit and pull the diamond from the base metal or fill voids in the dust pattern with material. The W-Point carbide drill design lasts longer and helps solve fuzz, delaminating, and burn problems when drilling.

2. What is the minimum edge distance allowed for aluminum alloy single lap sheet splices containing a single row of rivets as compared to a joint with multiple rows, all rivets being equal in diameter?

A) The minimum edge distance for the single row is greater than that for the multiple row.

B) The minimum edge distance for the single row is less than that for the multiple row.

C) The minimum edge distance for the single row is equal to that for the multiple row.

Answer: C. Learning Statement: AMG004. AC 43.13-1B, Acceptable Methods, Techniques, and Practices—Aircraft Inspection and Repair.

The minimum edge distance is to be not less than two times the diameter of the rivets used for both single and multiple row single lap sheet splices.

3. What is commonly used to connect an emergency source of power, and at the same time disconnect the normal hydraulic source from critical parts of a landing gear or wheel braking system for operation (usually when the normal source system fails)?

A) Selector valve.

B) Shuttle valve.

C) Sequence valve.

Answer: B. Learning Statement: AMA063. AMR, Aircraft Maintenance and Repair.

The function of a shuttle valve is to provide a means of disconnecting a normal source of hydraulic (or pneumatic) power and connecting an emergency source of power (hydraulic or pneumatic) to operate the critical parts of a system.

Aviation Maintenance Technician—Powerplant

1. If an un-supercharged reciprocating engine equipped with a constant speed propeller is operated at part throttle and at cruising rpm, a reduction in rpm with no change in throttle setting will result in

A) no change in manifold pressure.

B) an increase in manifold pressure.

C) a decrease in manifold pressure.

Answer: B. Learning Statement: AMP070. JSPT, A&P Technician Powerplant Textbook.

A reduction in rpm setting (propeller pitch increase) on an unsupercharged reciprocating engine equipped with a constant speed propeller, with no change in throttle setting, will cause an increase in manifold pressure. In this case, the decrease in rpm is caused by a higher load being placed on the engine rather than a reduction in fuel flow into the engine.

2. What are the two main sections of a turbine engine for inspection purposes?

A) Hot and cold.

B) Combustion and exhaust.

C) Compressor and turbine.

Answer: A. Learning Statement: AMP069. JSPT, A&P Technician Powerplant Textbook.

For inspection purposes, the two main sections of a turbine engine are hot and cold. The cold section includes the compressor back through the diffuser. The hot section includes the combustor and turbine.

3. Aluminum propeller blade failure at the site of an unrepaired nick or scratch is usually the result of

A) material defect.

B) intergranular corrosion.

C) stress concentration.

Answer: C. Learning Statement: AMP052. AP, Aircraft Powerplants.

Even a small defect such as a nick or scratch causes a concentration of stresses that may develop into a crack. The crack in turn results in even greater stress concentration. The resulting growth of the crack will almost inevitably result in blade failure.

TECHNICAL REFERENCES

The publications listed on the following pages contain study material you need to be familiar with when preparing for aviation maintenance technician knowledge tests. All of these publications can be purchased through U.S. Government bookstores, commercial aviation supply houses, or industry organizations. The latest revision of the listed references should be requested. Additional study material is also available through these sources that may be helpful in preparing for aviation maintenance technician knowledge tests. All publications listed would be excellent for a mechanic to have in a personal reference library.

The following lists contain abbreviations used to identify the reference(s) associated with a specific test.

Aviation Maintenance Technician—General

AMT-G	Aviation Maintenance Technician Series General—Aviation Supplies & Academics (ASA), Inc.
ABS	Aircraft Basic Science—Glencoe/McGraw-Hill
AP	Aircraft Powerplants—Glencoe Division, Macmillan/McGraw-Hill

AEE	Aircraft Electricity and Electronics—Glencoe/McGraw-Hill
AC	Advisory Circular—Federal Aviation Administration (FAA), Government Printing Office (GPO)
AIM	Aeronautical Information Manual—FAA, GPO
14 CFR	Title 14 of the Code of Federal Regulations (part or section (§))—GPO
FAA-G-8082	Guide—FAA, GPO
FAA-H-8083	Handbook—FAA, GPO
MBM	Marathon Battery Instruction Manual
ECD	Electronic Circuit Devices—Jeppesen Sanderson, Inc.
AB	Aircraft Batteries, Lead Acid/Nickel-Cadmium—Jeppesen Sanderson, Inc.
ATD	Aircraft Technical Dictionary—Jeppesen Sanderson, Inc.
JSGT	A&P Technician General Textbook—Jeppesen Sanderson, Inc.
JSPT	A&P Technician Powerplant Textbook—Jeppesen Sanderson, Inc.

Aviation Maintenance Technician—Airframe

AC	Advisory Circular—Federal Aviation Administration (FAA), Government Printing Office (GPO)
AEE	Aircraft Electricity and Electronics—Glencoe/McGraw-Hill
AMR	Aircraft Maintenance and Repair—Glencoe/McGraw-Hill
AP	Aircraft Powerplants—Glencoe/McGraw-Hill
AMT-A	Aviation Maintenance Technician Series Airframe—Aviation Supplies & Academics (ASA), Inc.
DAT	Dictionary of Aeronautical Terms—ASA
AAC	Aircraft Air Conditioning (Vapor Cycle)—Jeppesen Sanderson, Inc.
FMS	Aircraft Fuel Metering Systems—Jeppesen Sanderson, Inc.
AHS	Aircraft Hydraulic System—Jeppesen Sanderson, Inc.
AOS	Aircraft Oxygen Systems—Jeppesen Sanderson, Inc.
JSAT	A&P Technician Airframe Textbook—Jeppesen Sanderson, Inc.

JSGT	A&P Technician General Textbook—Jeppesen Sanderson, Inc.
ABS	Aircraft Bonded Structure—Jeppesen Sanderson, Inc.
WG	Welding Guidelines with Aircraft Supplement—Jeppesen Sanderson, Inc.
ARS	Aircraft Radio Systems—Jeppesen Sanderson, Inc.
AComp	Advanced Composites—Jeppesen Sanderson, Inc.
14 CFR	Title 14 of the Code of Federal Regulations (part or section (§))—GPO
47 CFR	Title 47 of the Code of Federal Regulations (part or section (§))—GPO
49 CFR	Title 49 of the Code of Federal Regulations (part or section (§))—GPO
FAA-H-8083-21	Rotorcraft Flying Handbook—GPO
MBM	Marathon Battery Manual
MMM	Manufacturer's Maintenance Manual
TSO	Technical Standard Order
SUND	Sundtrand IDG and BITE 767 Line Maintenance/Servicing

Aviation Maintenance Technician—Powerplant

ABS	Aircraft Basic Science—Glencoe/McGraw-Hill
AC	Advisory Circular
AEE	Aircraft Electricity and Electronics—Glencoe/McGraw-Hill
AMR	Aircraft Maintenance and Repair—Glencoe/McGraw-Hill
AMT-G	Aviation Maintenance Technician Series General—Aviation Supplies & Academics (ASA), Inc.
AMT-P	Aviation Maintenance Technician Series Powerplant—ASA
AP	Aircraft Powerplants—Glencoe/McGraw-Hill
DAT	Dictionary of Aeronautical Terms—ASA
TCAS	Transport Category Aircraft Systems—Jeppesen Sanderson, Inc.
APC	Aircraft Propellers and Controls—Jeppesen Sanderson, Inc.
ATD	Aircraft Technical Dictionary—Jeppesen Sanderson, Inc.
JSGT	A&P Technician General Textbook—Jeppesen Sanderson, Inc.

JSPT	A&P Technician Powerplant Textbook—Jeppesen Sanderson, Inc.
AGTP	Aircraft Gas Turbine Powerplants—Jeppesen Sanderson, Inc.
14 CFR	Title 14 of the Code of Federal Regulations (part or section (§))—Government Printing Office (GPO)
PSG	A&P Technician Powerplant Study Guide—Jeppesen Sanderson, Inc.

Notes:

Notes:

Notes:

Notes:

Notes:

Made in United States
Troutdale, OR
11/03/2026

56844463R00015